The Problem with Education Technology

(Hint: It's Not the Technology)

The Problem with Education Technology

(Hint: It's Not the Technology)

Ben Fink

Robin Brown

Utah State University Press

Logan

Published by Utah State University Press
An imprint of University Press of Colorado
5589 Arapahoe Avenue, Suite 206C
Boulder, Colorado 80303

USU Press Current Arguments in Composition

The University Press of Colorado is a proud member of
The Association of American University Presses.

The University Press of Colorado is a cooperative publishing enterprise supported in part, by Adams State University, Colorado State University, Fort Lewis College, Metropolitan State University of Denver, Regis University, University of Colorado, University of Northern Colorado, Utah State University, and Western State Colorado University.

ISBN 978-160732-446-1 (pbk)
ISBN 978-160732-447-8 (ebook)
DOI: 10.7330/9781607324478

Library of Congress Cataloging-in-Publication Data

Names: Fink, Ben, 1984– | Brown, Robin, 1944 May 22–
Title: The problem with education technology : (hint—it's not the technology) / Ben Fink, Robin Brown.
Description: Logan : Utah State University Press, [2015] | Includes bibliographical references.
Identifiers: LCCN 2015012982 | ISBN 9781607324461 (pbk.) | ISBN 9781607324478 (ebook)
Subjects: LCSH: Examinations—United States—Scoring—Data processing. | Grading and marking (Students)—United States—Data processing. | Educational tests and measurements—United States—Data processing. | Educational technology—United States.
Classification: LCC LB3060.83 .F55 2015 | DDC 371.33—dc23
LC record available at http://lccn.loc.gov/2015012982

Contents

The Problem with Education Technology (Hint: It's Not the Technology)

TEACHERS VS. TECHNOLOGY—ROUND ONE, FIGHT!

> *In today's society, college is ambiguous. We need it to live, but we also need it to love. . . . Teaching assistants are paid an excessive amount of money. The average teaching assistant makes six times as much money as college presidents. In addvition, they often receive a plethora of extra benefits such as private jets, vacations in the south seas, a staring roles in motion pictures. Moreover, in the Dickens novel Great Expectation, Pip makes his fortune by being a teaching assistant.*[1]

It was our last day at the 2012 Conference of College Composition and Communication (CCCC) in St. Louis, the biggest annual gathering of college writing teachers in the country. We were done with our own presentation, and we could eat only so many ribs in a day . . . so we were looking through the conference program to see what else was on offer . . . when a full-page listing caught our eye. It was for a "Special Session." Something particularly Big and Important. This Special Session, the program told us, would be a "vigorous but civil debate" on Automated Essay Scoring (AES), the technology that lets computers score writing assignments.

We were intrigued. When an academic conference has to specify that an event will be "civil," you know something's up.

The ballroom was vast and long. There were about two hundred seats facing a long table and a giant PowerPoint screen. At one end of the table sat Paul Deane and Chaitanya Ramineini, researchers from Education Technology Services (ETS): the College Board, the minds behind standardized tests like the SAT, GRE, and AP exams, and—along with their for-profit colleagues Pearson and Vantage—the loudest promoters of AES. They were here today to present their newest innovation: the e-Rater® Scoring Engine, a computer program that—according to an impressive set of studies—can score a student essay with the reliability of a human.

At the other end of the table sat Les Perelman, the longtime (now emeritus) Director of Undergraduate Writing at MIT—and the best-publicized crusader against AES. He authored the first paragraph of this section, which he fed, along with another page-and-a-half of similar drivel, into e-Rater®. And it got a 6, the highest-possible score. The point seemed clear and obvious: machines can't read. They can't understand. And they can't, and certainly shouldn't, replace humans in educating our children.

(Between them, fittingly, sat Carl Whithaus, director of the University Writing Program at UC Davis, who works to integrate AES into human-centered teaching. We'll get back to him later.)

Les was the perfect foil to Deane and Ramineini. They were impeccably dressed; he wore a suit that didn't fit quite right and looked a little sweaty. Their speech was memorized and newscaster-perfect; he spoke off the cuff and made no effort to hide a general-purpose ethnic accent. Their PowerPoints were sleek, branded, and serious; his was homemade and funny, complete with gifs of rabbit holes and the *Twilight Zone*.

None of this was by accident. They were each playing their parts to a T: the bloodlessly efficient technocrats versus the righteous, rumpled, lone defender of flesh-and-blood teaching. "Vigorous but civil" be damned—this was a brawl. An *agon*. A pundit war between good and evil, man and machine. We, the human educators of CCCC, sat and winced as the forces of evil took their best shot—and then rejoiced as our champion tore them limb from limb, down to their last unfounded assumption and logical fallacy.

He was, after all, fighting for our lives. Or at least our livelihoods. Humanistic objections to robo-teaching aside, we were precisely those humans AES stands poised to eradicate. AES, which can grade thousands of student essays in mere seconds, could convince a budget-wary public that small class sizes and individualized instruction are unnecessary luxuries. In a political moment where education funding is under constant attack, it's not hard to imagine administrators and elected officials (even the well-meaning ones) using AES as a rationale to lay us off. So when we see Les up there proving that AES doesn't work, won't work, can't work, of course we cheer loudly.

The problem is, outside the cozy confines of CCCC, he's losing. We're losing. Education technology—computer-based classroom instruction, online K–12 schools, MOOCs (Massive Open Online Courses), as well as AES—is all the rage. It's what's getting all the media attention, foundation funding, and government grants. It's the cutting edge, the thing all forward-thinking principals, superintendents, policymakers, and executives (both public and private) are talking about. Les's dogged defense of old-fashioned teaching might make us cheer, but it's making everyone else yawn.

And maybe most frustrating of all, the science all seems to be on their side. Sure, Les's argument makes intuitive sense—

how could you trust a program that gave that pile of s#!t he wrote the highest-possible score?—but then again, how can you argue with scores of reliable, data-driven studies?

This is the problem we'll tackle in this book. We'll explain the technology, how it works, and what it does (and doesn't do). We'll explain where it comes from, and how it's come to take the form that it has, in the classroom and in our culture. We'll also explain why the debate over the technology has taken the shape it has—a shape that stops us from understanding the real problem or doing anything about it. Finally, we'll explain what that real problem is, and how we—teachers, students, citizens—can work to solve it.

Because there is a problem. It's just quite a bit subtler, tougher, and more complicated than the standard "civil debate" would have us believe.

THE FIGHT OVER COMPUTER GRADING (STARTED BEFORE COMPUTER GRADING)

Human versus machine. Good versus evil. Teachers versus technology. Kids versus computers. This is a Big Issue, and not just for writing teachers.

A few weeks after the conference, it made the *New York Times*. "Facing a Robo-Grader? Just Keep Obfuscating Mellifluously," read the headline. The story, by longtime *Times* reporter Michael Winerip, starts with a recent study that "concluded that computers are capable of scoring essays on standardized tests as well as human beings do"—at the rate of 48,000 essays per minute.

It gives some space to ETS representatives, including Paul Deane. But most of the story belongs to Les, who demolishes the software with his usual gusto:

> *The e-Rater's biggest problem, he says, is that it can't identify truth. He tells students not to waste time worrying about whether their facts are accurate, since pretty much any fact will do as long as it is incorporated into a well-structured sentence. "e-Rater doesn't care if you say the War of 1812 started in 1945," he said.*
>
> *Mr. Perelman found that e-Rater prefers long essays. A 716-word essay he wrote that was padded with more than a dozen nonsensical sentences received a top score of 6; a well-argued, well-written essay of 567 words was scored a 5.*
>
> *An automated reader can count, he said, so it can set parameters for the number of words in a good sentence and the number of sentences in a good paragraph. [. . .] e-Rater likes connectors, like "however," which serve as programming proxies for complex thinking. Moreover, "moreover" is good, too.*
>
> *Gargantuan words are indemnified because e-Rater interprets them as a sign of lexical complexity. "Whenever possible," Mr. Perelman advises, "use a big word. 'Egregious' is better than 'bad.'"*[2]

The ETS representatives make some important points in response, which we'll get to later. But they don't—and can't—refute Les's basic claims. Computers can't read. They can't understand truth. They can only grade by counting things: length of essays, length of sentences, and length of words. So, obviously, they shouldn't be used in place of humans. Right?

Not so fast. Take a look at this:

> *In March, Les Perelman attended a national college writing conference and sat in on a panel. [. . .] "It appeared to me that regardless of what a student wrote, the longer the essay, the higher the score," Dr. Perelman said. A man on the panel from the College Board disagreed. "He told me I was jumping to conclusions," Dr. Perelman said. "Because M.I.T. is a place where everything is backed by data, I went to my hotel room, counted the words in those essays and put them in an Excel spreadsheet on my laptop. [. . .]*

> *He was stunned by how complete the correlation was between length and score. "I have never found a quantifiable predictor in 25 years of grading that was anywhere near as strong as this one," he said. "If you just graded them based on length without ever reading them, you'd be right over 90 percent of the time." The shortest essays, typically 100 words, got the lowest grade of one. The longest, about 400 words, got the top grade of six. In between, there was virtually a direct match between length and grade. He was also struck by all the factual errors in even the top essays. [. . .]*
>
> *How to prepare for such an essay? "I would advise writing as long as possible," said Dr. Perelman, "and include lots of facts, even if they're made up." This, of course, is not what he teaches his M.I.T. students. "It's exactly what we don't want to teach our kids," he said.*[3]

This is another story from the *New York Times*. Like the first one, it's about a fight between Les and ETS at CCCC. Like the first one, Les is criticizing ETS for scoring essays based on length, rather than truth or quality. And like the first one, it's pretty darn convincing.

Only this story isn't from March 2012. It's from March *2005*. And it wasn't about AES at all.

This story was, rather, about the then-brand-new essay portion of the SAT. (Which is now defunct, due in no small part to Les's efforts.) It seems the SAT essay had all the same problems Les now identifies with AES. But computers didn't grade these essays.

Here's where the plot thickens. Yes there's a problem with the technology, but it's not just about the technology—take away the computers, and the problem persists.

To understand why, first we need to understand more about how the computers work, how they're programmed, and what, exactly, they're programmed to do.

HOW THE TECHNOLOGY WORKS

How does a computer grade an essay? If you believe the hype, including some tech companies' marketing materials, AES works pretty much like HAL in *2001*: it passes the Turing test. It can talk, think, read, and respond just like a human. It "understands the meaning of written essays."[4]

That's a pretty big claim. And it's false. There is indeed a fruitful and fascinating line of research pointing to this kind of "total" Artificial Intelligence (AI): a theory of human cognition and language that could model the near-impossibly complex workings of the mind, as meanings are made, communicated, and understood. But so far, it's purely academic, with pretty much no practical application. We know a whole lot more about the structures of the brain, the mind, and language than we used to, but we're still a long way from a machine that really "understands meaning."

So what *do* AES systems do? Professor Patricia Freitag Ericsson explains:

> *These programs treat essays as pure information that can be mined for some abstracted set of words that, at least to their promoters, equates to meaning. The shifting, dynamic relationships that these words have to each other, to society, and to different readers is invisible to these information-seeking machines. The machines can tell users whether writers have matched the words in an essay with words in a database (or a triangulated database matrix), but they cannot assess whether this mix of words conveys any meaning.*[5]

The people who make the AES technology know this full well. When they're writing seemingly-scholarly articles, rather than doing PR, they fully own up to it. Here's Dr. Randy Elliot Bennett, the "Norman O. Frederiksen Chair in Assessment Innovation" at ETS:

> *It is important to note that, regardless of what vendors may claim, a machine does not read, understand, or grade a student's essay in the same way as we like to believe that a human rater would. It simply predicts the human rater's score.*[6]

How does it do this? By using a second, distinct, and far less ambitious form of AI: an *expert system.* Expert systems are meant to *emulate the external behaviors* of flesh-and-blood experts. Not to *be* experts, or even to try, just to act like them in the most surface-level way. Give them the same inputs you'd give a human expert, and they'll produce the same outputs. (The best-known example of an expert system is Watson, the machine that beat Ken Jennings at *Jeopardy*, and now does medical diagnosis.)

How do you build an expert system? It's pretty simple, actually:

1. *Define a discrete phenomenon, identifiable by human experts given certain data.* Like a disease, which doctors can identify based on a bundle of symptoms; or "good writing," which teachers can identify based on a bundle of traits.
2. *Assemble a body of experts.* Load up the program with terabytes of examples from lots of real human experts—reams and reams of charts, test data, notes, etc.
3. *Describe and segment the behaviors involved in "expertise."* Basically Taylorism, twenty-first century-style: break down the work human experts do into as many discrete parts as possible, and teach the computer program to imitate each of those parts.
4. *Define the "markers" that emerge as significant identifiers.* Those patterns of symptoms, test results, traits, etc. that lead human experts to reach one conclusion versus another.

5. *Rank and relate these markers to each other*. In order to emphasize the ones that human experts rely on the most, the ones that "really matter."
6. *Test and repeat*. And intelligent systems, like Pandora, will "learn" from experience.

That's it. Expert systems are not independently smart. They can't replace human experts, or even "think" like them. And they're not trying to. All they do is match input data with output conclusions, based on all the past examples programmed into them—they match the evidence the human experts saw with the conclusions they reached based on that evidence.

THE TECHNOLOGY'S "EXPERTISE" (IS NO MORE LIMITED THAN THAT OF ITS HUMAN COUNTERPARTS)

So: AES doesn't "read" essays, let alone "understand" them. It just uses its algorithms to predict the behaviors of certain human "experts."

And these "experts" are, frankly, rather predictable. We're not talking about English teachers pouring lovingly over the work of their twenty students. We're talking about the kind of human "expertise" that ETS and its many, many fellow testing companies have always relied on: standardized-test raters, sitting in a room, reading hundreds of formulaic essays at breakneck speed and assigning points based on the presence of a few formally-defined "primary traits." The same essay scorers Les criticized, rightly, in the *Times* story from 2005.

If you are under the age of sixty and grew up in the United States, you've probably taken some kind of standardized essay test. You may (but probably don't) remember the prompt. It was vaguely moral and social issue-focused—like what Emerson might have encountered at Harvard back in the day. Here's a recent example:

> *Identifying with a group makes people feel secure with and trust one another because of what they have in common. They might share the same interests, language, beliefs, ethnicity, or cultural background. However, by limiting their identities to a specific group, people may miss important opportunities to connect with and understand others.*
>
> *Assignment: Is a strong group identity a good thing or a bad thing? Plan and write an essay in which you develop your point of view on this issue. Support your position with reasoning and examples taken from your reading, studies, experience, or observations.*[7]

We'll bracket some very important questions here. (Is this prompt fair to students of various ethnic/racial/gender/class backgrounds? And does anyone—test-makers, test-takers, test-scorers—actually care about the prompt, or the answers being offered?)

The question we need to ask is: once thousands and thousands of students have tried to answer this prompt in twenty-five-minute timed writing tests, and thousands of teachers have gathered for a weekend in a hotel to score them all (working to supplement their meager salaries with summer work), how do they do it?

Quickly. Very quickly. In ETS scoring sessions, the model all other companies follow, the readers work like maniacs, reading against the clock. No more than two to three minutes per essay. (They also take periodic breaks for "re-norming" sessions, where they compare each other's ratings and work toward the "inter-rater reliability" that ETS prides itself on—the same consistency of input-to-output that the AES software strives for.)

Within those two to three minutes, what are they looking for? The basic ETS scoring rubric has remained the same for decades. It has five categories, and raters assign a score of 0–6 for each. Here they are, slightly paraphrased:

1. Point of view, evidence, and support.
2. Organization and coherence—as marked by transition and linking words.
3. Vocabulary.
4. Sentence complexity.
5. Grammar, usage and mechanics.[8]

What's most striking is what's missing. Nothing about getting a complex idea into words. Or presenting big, bold claims in convincing ways. Or crafting language, stories, and arguments that can move, delight, or agitate an audience. Or any other things that (we think you'll agree) actually make for good writing.[9]

Good writing, here, is just about *form*—in the simplest and silliest of ways. It's about having big words, complex sentences, no "grammar" errors, lots of "furthermores" and "moreovers," and maybe some sense of a topic you can stick to. The kind of stuff you can spot from a mile away—or at least, a few feet:

> *Dr. Perelman is now adept at rapid-fire SAT grading. This reporter held up a sample essay far enough away so it could not be read, and he was still able to guess the correct grade by its bulk and shape. "That's a 4," he said. "It looks like a 4."*[10]

Les Perelman really *is* an expert essay-evaluator, and AES expert-system programs do what he does. Which, as we see, is not much. Recall ETS's Dr. Bennett: "a machine does not read, understand, or grade a student's essay in the same way as we like to believe that a human rater would." The key phrase is *we like to believe*. A machine cannot read a student's writing in the way *we like to believe* that a human rater would. But it *can* respond in the way human raters *actually do*.

The problem with the machine isn't about the machine. It's about how humans (in this case, standardized test raters) got turned into machines. And how these mechanized humans become accepted as the "experts" for the machines to (easily) emulate. And ultimately, how a set of silly, superficial, fakeable formal features became synonymous with "good writing"—how schoolwork itself became something mechanical.

THE ENDURING LURE OF LABOR-SAVING DEVICES

> *Imagine taking a college exam, and, instead of handing in a blue book and getting a grade from a professor a few weeks later, clicking the "send" button when you are done and receiving a grade back instantly, your essay scored by a software program.*[11]

Here's the start of another recent *New York Times* story, by technology reporter John Markoff. This one's a little more optimistic. It announces the new "nonprofit enterprise" EdX. EdX, along with similar enterprises like Coursera® and Udacity, is working to turn courses from elite universities around the world into MOOCs. Massive Open Online Courses, available not quite for free, but certainly for a lot less than a college tuition.

These MOOCs, Markoff explains, would use AES software to handle their huge enrollments. Though "skeptics say the automated system is no match for live teachers"—he cites a group of dissenting professors led by a certain Dr. Perelman—"EdX expects its software to be adopted widely by schools and universities." The size, scale, and nationwide demand for MOOCs, he concludes, will make AES indispensable.[12]

By now, you'll find this story familiar. Once again we have—in this corner—the new, innovative "enterprises" and the international cast of engineers, administrators, and political boosters trumpeting the wondrous possibilities of our brave new technology-driven world. And in that corner, always presented in reaction to the sunny engineers, are the "skeptics," the Debbie Downers, the crotchety old professors who insist the technology can't work and would you please get off my lawn, inevitably led by Les Perelman.

As always, the skeptics win the arguments, but the enterprises win the hearts and minds—and the contracts. It's just the way the world is moving, we're told. It may have its problems, sure, but There Is No Alternative.

The only part of the story that *wasn't* familiar was the headline. "Essay-Grading Software Offers Professors a Break." Well then! With all the teacher-bashing talk that dominates our national conversation about education, it's not every day we teachers get offered a break!

Not likely. The headline was probably written by a hasty editor; Markoff never discusses it. Yet however unintentionally, it points us to a crucial and overlooked issue. Maybe *the* crucial and overlooked issue. Namely this: AES, like most hyped education technologies, is a *labor-saving device*. That's its whole purpose: to save humans the labor of reading and evaluating student writing—either to give them a break, or more likely these days, to replace them (and save a lot of money).

AES and MOOCs may be new, but labor-saving devices are anything but. Pretty much every major new education technology of the past half-century has been a labor-saving device, either meant to make teachers' lives easier or to make teachers obsolete.[13]

In either case, the technology is responding to a very real problem: there isn't enough labor to go around. A teacher with thirty-five students per class—which is to say, over 150 at a time—really *can't* devote the necessary labor to each student. Especially when they're in underfunded classrooms, with students from lots of different cultures speaking multiple languages, and often struggling with poverty, hunger, and violence.

The enduring lure of the technology is that it can fix this problem. It can do what the overworked, under-resourced teacher can't. It can provide each and every student with the individualized instruction, the caring feedback, and the reliable support that he or she needs to succeed, all within the ever-shrinking budget. It can ensure that, in fact, No Child will be Left Behind.

But can it?

(HOW) DO LABOR-SAVING DEVICES WORK?

In a word: no. At least, not as advertised.

This shouldn't be controversial. After decades of new technologies, each promising to fix everything, the same problems still persist.

Yet we fall for it, every time. Sure, everything up until now has failed to save the children, but *this new piece of technology, this new app, this new program, this new labor-saving device* will really change things.

It won't. Just like all the others haven't. We're not saying this because we're grouchy Luddites. (We like technology, and we use a lot of it in our teaching.) It's just that structurally speaking, technology *used as a labor-saving device* doesn't work to educate people. Because getting educated is, precisely, about labor.

How do we know? Just take a look at who's getting left behind, and who's not.

The kids who succeed, now just as much as a half-century ago, are the ones who have had a lot of labor invested in them. In school: in the form of small classes, tons of teacher interaction, and ample opportunities for extracurricular enrichment (the debate team, the jazz band, the Intel Science Talent Search). And just as importantly, outside of school: in the form of well-resourced families and communities.[14]

Just think for a second about how you got educated. If you're anything like us, you had some good teachers, but you also had a lot more. The two of us have somewhat differing experiences. One of us (Ben) had a pretty typical upper-middle-class upbringing: two professional parents who encouraged their kids to express themselves, art and music lessons, summer camp, travel, and a house in a well-off suburb with parks and kids playing outside and a well-funded public school system that gave students lots of attention.

The other (Robin) has a less likely story: lots of alcohol and its familiar outcomes, but however confused the family talk may have been, it was smart. There were novels around the house, and the *Saturday Evening Post* and *National Geographic* showed up in the mail. So did *Popular Mechanics*—whose wacky, futurist ideas could often be turned into reality in the shop. And there was a caring mentor to teach him to weld and solve mechanical problems, and to talk about ethics and politics—and to give him a concrete image of himself as competent and smart.[15]

What's important is what these stories have in common. Both of us, in our different ways, had a whole lot of labor invested in us. As Hillary would have it, it took a village.

Or in the language of sociologist Pierre Bourdieu, it took a lot of *cultural capital*. All those conversations around the dinner table, all those hours in the woodshop, all those teachers spending one-to-one quality time mentoring us, all of that was an *investment*. Specifically, an investment of labor, in training our bodies to move comfortably in particular ways.

We mean this literally. Many of us, now, can get into a car and drive off—without ever thinking about it. But think about the first time you tried. It was really awkward. (Which pedal do you push? How hard? (Wait, what's that scraping sound?)) It only came to feel "natural" through a huge investment of labor—lots of practice, and maybe driver's ed.

Same with pretty much *everything* we've learned to do "naturally." Think about manners. We may not remember getting taught how to sit at a table, make eye contact, nod appropriately as people talk—but we did. And we immediately know when we're in a situation where we don't have the "right" manners, because we get fidgety, and suddenly we don't know what to do with our hands.

Same goes for schoolwork. To be able to sit down and write an "effective" and "insightful" essay on a 25-minute standardized test, you need a body (not just a brain) that's been well trained in a highly specific way. Knowing how to move "naturally" through a school assignment is a lot like knowing how to hold a glass of champagne in one hand and a piece of bread with no crust and three capers in the other, all the while commenting on the delicate brushwork and color palette of the Monet on the wall, and making it all look effortless.

And it's no less class-based. The ability to move well, at a cocktail party or in a classroom, is an example of what Bourdieu calls "durable dispositions." Dispositions, because they're the ways our bodies feel naturally disposed to move. Durable, because they're hard to change—hard to get, and hard to get rid of.[16]

Different dispositions have different value in different places. (Or in Bourdieu's terms, on different *fields* or *markets*.)[17] The kinds of dispositions that'll get you a score of 6 from ETS (human rater or e-Rater®) won't get you a score of 10 at a poetry slam—or vice versa. Both kinds of dispositions are equally valid and valuable, at least in our opinion. But only one will get you into college.

There's no way to get those dispositions, other than investing a lot of labor. That's why labor-saving devices haven't worked, don't work, can't work. If your body doesn't have the dispositions, it can't act as if it did.

And when it tries, it fails.

FAKING IT: WHY RICH KIDS CAN DO IT AND POOR KIDS CAN'T

What can a poor student do? One who doesn't have the necessary dispositions but has to take standardized tests anyway?

Go to actoutagainstsat.com—designed by some savvy (and affluent) high school students who don't much like the SAT—and you'll find a link to a document called "Dr. Perelman's SAT Writing Tips." ("Maybe it is wrong to give out tips for something so flawed," the students explain, "but until we can get rid of the tests, why not help everyone get higher scores?")

Dr. Perelman's tips include the following:

> *Before the test, prepare two historical biographies and two works of literature as examples. If the examples don't fit, change them so they do. Details count; factual accuracy doesn't.*
>
> *Use "plethora" and "myriad" along with other "big" words. Don't be afraid that you are using the words incorrectly. Use of "big" words counts!*
>
> *Have two or three quotations ready. Insert one in the concluding paragraph, even if it isn't really relevant to your thesis or your argument. My favorite is FDR's "the only thing we have to fear is fear itself."*[18]

In a word (or two): fake it.

Does this kind of thing actually work? In some cases, yes. When we met Les at CCCC a few years ago, he told us some (hilarious) success stories from when his students went into the Boston Public Schools to teach these tips. But he didn't tell us much about who those successful test-takers were. If past experience serves, the ones who could successfully fake it were the ones who were already pretty much able to do it for real. (That is, they already had the necessary dispositions.)

There are a lot of fakes out there. In fact, it's a whole industry.

Have you looked at SparkNotes lately? If not—if the name means nothing to you, or if you haven't seen it since back when it was basically online Cliffs Notes™ (which Ben used to not-read *Madame Bovary* in AP English)—go take a look. It's a slick, Flash-y social media site. Movie trailers. Celebrity gossip. Emma Watson talking about feminism. All shareable on Facebook.

The school-related content is actually pretty good. The fake blog we saw on *The Grapes of Wrath*—advertised with a picture of a cute, beleaguered student in front of a laptop—asks real, probing questions about the novel and fakes a few good responses. And then you can see where real kids signed on, offered their own (insightful) thoughts, and engaged in a fairly high-level discussion of the novel.

Of course, it's still primarily a cheating site. But nothing about the site encourages its users to think of themselves as "cheaters." No, it implies, they're good students. Sleek, solid, successful students. Probably headed for Stanford. Far too sophisticated to waste time actually doing the reading.

SparkNotes can be a very efficient labor-saving device—if you know how to use it. If you're in the "right" kind of school, your peers will lead you to it and show you how. But like Les's SAT tips, you've got to be solidly literate to be able to use it

. . . without getting caught. Schools these days are obsessively on the alert for "plagiarism." Cut and paste wholesale from SparkNotes into an assignment, and you're screwed. But read it, excerpt it, and fill it in with your own homemade fakery, and you'll probably do fine. It's the rare teacher who'll have time to spot what came from the for-profit, monetized, data-mining website, and what didn't.

The same class divide is on display in the Princeton Review's $1,000+ SAT test prep course. (Ben took it; Robin didn't.) The strategy they teach is centered on a fictional character they call Joe Bloggs: the average, predictable student with average, predictable dispositions, who falls into all the test's carefully-laid traps. The Princeton Review's basic lesson, woven through all their classes, is: "Don't be Joe." They teach you how, in minute detail. And it works: they guarantee a 150-point increase, or your money back.

Take away all the money, and all the finely-honed dispositions it buys, and the fakes stop working. No one makes this clearer than Ed Burns, veteran Baltimore Public Schools teacher and co-producer of HBO's *The Wire*. He shows us Mr. Prezbo, a well-meaning but naive [white] cop-turned-schoolteacher forced to prepare his [mostly poor African American] students for high-stakes standardized testing. In one scene, we see him trying to teach an eighth grade class to fake the writing sample. He tells them to start their essays—regardless of the prompt—with "According to the text," and end with "The above stated facts have led me to conclude that . . ."

The students don't get it. They have no context for any of this; they don't even understand what they're trying to fake. And come test time, they do horribly.

This is pretty much ETS's response to Les's SAT tips. Again according to Michael Winerip in the *New York Times*:

> *E.T.S. officials say that Mr. Perelman's test prep advice is too complex for most students to absorb; if they can, they're using the higher level of thinking the test seeks to reward anyway. In other words, if they're smart enough to master such sophisticated test prep, they deserve a 6.*[19]

They're mostly right. You can't fake it unless you've already actually got it. There's no way to get a 6 without actually having all the dispositions you need to get a 6. Only it's not about being "smart enough," as ETS claims. It's a matter of being *privileged* enough: lucky enough to have had all the necessary labor invested in you. If you've got it, you'll be able to do it—or, if you prefer, fake it. If you don't, you won't—and no labor-saving device will save you.

THE MATTER WITH MOOCs

When we got back from CCCC in 2013—the same year we talked with Les—we both found the same email in our inboxes, from someone named Joe Moxley, a professor at the University of South Florida. We asked a few colleagues, and they confirmed our suspicion: Professor Moxley had spammed the entire CCCC national network, to announce that

> *Professor Denise Comer's team from Duke University launched its ground-breaking Composition MOOC, English Composition I: Achieving Expertise. In case you missed the announcement, Duke's MOOC is supported by the Bill & Melinda Gates Foundation and hosted by Coursera.*

MOOCs—those Massive Open Online Courses—are seductive. They dangle in front of us no less than the possibility of a global educational utopia. According to Coursera®'s website:

> *We envision a future where everyone has access to a world-class education that has so far been available to a select few . . . to*

> *empower people with education that will improve their lives, the lives of their families, and the communities they live in.*

The reality is a bit more complicated. MOOCs may be open and online and available to pretty much anyone, but they come up short on the education side. They lack any meaningful teacher-student interaction. They often lack substantial reading and writing assignments. Their dropout rate hovers around 90 percent.[20] Even Sebastian Thrun, the Silicon Valley entrepreneur who invented MOOCs, now admits they don't work.[21]

We shouldn't be surprised. All the hype aside, MOOCs are just another labor-saving device. They offer a lot of promise, but like all labor-saving devices, they don't work for the ones who need it most—in Thrun's words, "students from difficult neighborhoods, without good access to computers, and with all kinds of challenges in their lives. . . . It's a group for which this medium is not a good fit."

And how about the other students? The actual Duke students, the ones paying a quarter million dollars and arriving with the appropriate dispositions, for whom this medium might be a good fit? They don't take MOOCs. They can't even get credit for them. ("Duke MOOC" is, in fact, a misnomer. Duke doesn't own it. Go to the Coursera® site, and you'll find the Duke brand prominently displayed. But go to Duke's site, and you won't see the Coursera® brand anywhere—or any mention of MOOCs at all.)

Duke students take the exact opposite of MOOCs: in-person writing classes, in groups of no more than twelve, taught by a faculty of about thirty-five postdocs from top schools all over the world. The instruction is strikingly old-school: just teachers doing intimate, face-to-face work with students, building relationships, and giving lots of feedback. Never mind that Duke University has every possible digital technology at its fingertips. They know what works. It's not MOOCs. Or AES. Or any labor-saving devices at all. Quite the opposite: it's a whole lot of labor.[22]

Which leads us to the obvious question: if MOOCs don't work for poor students, and rich students don't use them, why are they still around . . . and still growing? (As we write this, Coursera® offers 747 different MOOCs on its site.) Why is every department in every major university across the country being driven to put their classes online, ideally in MOOC form?

Simple. Labor-saving devices mean lower labor costs. Just like in the car plant: using more robots means paying fewer workers. Labor-saving devices may not work for students or teachers, but they work wonders for bureaucrats. And they're the ones making the decisions.

The history of education policy, at least over the past decade-and-a-half, is the history of cuts. Tax cuts, leading to spending cuts, leading to funding cuts, leading to budget cuts. And in this hostile, threatening environment, nothing looks better than a MOOC. It's new, it's high-tech, and it brings in grant money and media attention. Best of all, it seems to offer a win-win—a way to do "more with less," to offer cutting-edge education even with all the cuts.[23]

Put the cuts and the labor-saving devices together, and add the widespread belief that education technology has made a good education available to all—fueled by *Waiting for Superman* and the rest of the massive Gates Foundation propaganda campaign for school privatization—and it adds up to a pretty awful situation for students. Especially those on the lower end of the socioeconomic spectrum. According to all the tech boosters, they have every opportunity to succeed. (Cuts or no cuts, they have a world-class Duke education right in front of them!) So when they don't succeed (like the 90 percent who drop out), it must be their own fault.

To put it bluntly: when we let labor-saving devices take over education, we tacitly consent to a cultural and political system that seems to demonstrate the rightness of class inequality, over and over again. And by doing so, it cements that inequality in place. The rich are rich because they've earned it; the poor are poor because they deserve it.

THE PROBLEM WITH "PAPERS"

So what can we do about all this? As citizens, and as teachers?

It's tempting to answer: "nothing." Like a young Luke Skywalker, it's easy to feel like the people with the power to make these decisions are "all such a long way from here."

They're certainly doing their best to make us feel this way. Decisions to cut teaching staff and adopt new labor-saving devices (the two often go together) typically get made by government officials, upper-level school administrators, and industry "consultants" (read: lobbyists), in rooms without any teachers or students present. So a certain amount of hopeless despair is understandable.[24]

Then again, we may have more control than we think—for better (as we'll discuss in a bit), and also for worse. Why are robots poised to take over our teaching? And why has school-writing, like so much of education, been reduced to something that robots can handle? Because we've let it. ETS may be somewhat cynical in suggesting to teachers (in their marketing) that "we are committed to understanding the demands of the classroom" and "your requirements drive us to develop products and services," but they are not incorrect.[25] Protest as we might, we as a profession are still deeply invested, and complicit, in the robotic approach to schoolwork.[26]

Exhibit A: the "paper."

We all know the "paper." It's what we had to write in school, all the time. It's so universal, we barely recognize it as a distinct form of writing—but it is. Sometimes it's called an "essay," as in "Automated Essay Scoring," but that's inaccurate. If Michel de Montaigne submitted his classic essays to e-Rater®, or a human equivalent, he wouldn't do very well. (Unsubstantiated claims. Lack of citation. Sentence fragments.) Neither would the contemporary essays in *Harper's* or the *Atlantic*. (When's the last time you saw a published essay where the first paragraph started with a cliché and ended with a "specific thesis sentence"?)

No, "papers" aren't essays. Nor are they articles (scholarly or otherwise), or news stories, or editorials, or reviews, or lab reports, or annual reports, or letters, or blog posts, or PowerPoint decks—or any other kind of writing that exists outside captivity. They are, rather, an artificial contrivance. A mutant hatched by test-makers and education standard-setters, with the tacit and sometimes explicit consent of teachers, which has somehow taken over.

What is a "paper," exactly? Just like pornography: you know it when you see it. It follows the rough format of a high-school five-paragraph essay: intro paragraph with thesis statement, body paragraphs with supporting evidence, conclusion-restating thesis with moral lesson. And when they're bad, as they tend to be, they sound something like this:

> *In today's society the individuals that are not homophobic see homosexuality as a "whatever" type of thing as well.*
>
> *Unlike your stereotypical artist, who more than likely ended up killing themself due to the deep undertow of depression, Mondrian sang a different song and dance. Literally.*
>
> *Visual representation of women for sexual purposes may not be a conscious effort, but the passive female when seen through the eyes of the heterosexual male, is inherently venereal.*

> *Beowulf reigns for fifty years and, with time moving continuously forward, he grows old.*[27]

These are all examples of "Shit My Students Write" (shitmystudentswrite.tumblr.com, a kind of support group for frustrated writing teachers). Looking at these samples, few of us would argue with the fecal designation. But what is it, exactly, that makes them so bad? It's not just the bad grammar, poor word choice, and awkward phrasing. It's that they just sound *fake*. Like they could've come right out of Les Perelman's faux essay for e-Rater®. It's no wonder a robot can grade them. They're pretty robotic to begin with. Humans do not actually communicate this way—in speech or in (any other kind of) writing.

Why do students write this way? Just ask them—or think back to when you were one yourself. They write this way because they *have to*. In Bourdieu's language, it's (a bad approximation of) what's valued on the market called "school." And after a lot of repetition, it becomes a disposition. It gets into your body and comes to feel natural. The teacher (or the test) says "write a paper on the theme of gender in *The Great Gatsby*," and the well-trained student body responds, as if by instinct: "Throughout history, gender relations have moreover been a contested factor." They know it's fake. They'll tell you it's fake. But they do it, because it's what gets rewards.

But why do we, their teachers, reward this kind of fakery? Why do we assign "papers" in the first place? Well, some of us don't. Many of our colleagues are opening up writing in lots of exciting, unconventional, real-world ways, and they deserve much more attention than they get. (We'll give them some in a moment.) But for most teachers, the question never comes up. We assign papers because . . . that's just what you do. That's the expectation. That's what's in the standards, in the textbooks, on the test.

And besides, our colleagues say, isn't it important for kids to learn how to make an effective argument, defend it with evidence, and get it down on paper in the form of sentences and paragraphs?

Of course it is. We must teach these skills. They are absolutely essential if we want to have any kind of democratic, free society. (And they don't hurt when you're looking for a job, either.)

But "papers" don't actually teach those skills. In some ways, they do the opposite. For students who come into the classroom with the basic dispositions (who've had enough labor invested in them), assigning "papers" turns them into cynics, skilled at faking it and rewarded for doing so. Even worse, for those who come in *without* the basic dispositions, like Mr. Prezbo's kids, they learn that school has nothing to offer them: they can't succeed, and maybe they don't deserve to.

WHAT THEY GET RIGHT (AND WE GET WRONG)

"We are complicit. We are where we are because for a long time now we have been asking for it."[28]

With this confession, as sad as it is accurate, professor Richard Haswell concludes his history of AES. We teachers, he observes, have gone back and forth between embracing every new technology without much thought and rejecting or ignoring every new technology with just as little thought—depending, it seems, on the phase of the moon. And all the while, we have continued to assign "papers," deliver lectures, give multiple-guess tests, and otherwise teach in thoroughly mechanizable ways—without recognizing what we were doing.[29]

But even if we didn't recognize it, they did. ETS, Pearson, Vantage, and the rest of the industry all saw the opportunity, and they made their move. They weren't the ones who decided that education would be impersonal and mechanical. They just acted like good corporations: they analyzed the shape of the market and developed products they thought, correctly, would sell.

Had they found a market that placed value on social, communicative, citizenship-driven education—the kind many of us would prefer—they would no-doubt have developed some very good technology for that kind of education. But given what they actually found, they reasonably intervened with a modest proposal: since we've already made education robotic, why not just replace the costly humans with *actual* robots?

"During those years," Haswell observes, as these corporations were rapidly developing and deploying their products in grade schools, high schools, and colleges nationwide, the response from teachers was "barely audible."[30] So he set out to make some noise. In 2013, he and several colleagues launched the Human Readers petition (www.humanreaders.org), which lists all the standard anti-technology arguments (starting with "computers cannot 'read'") and "call[s] for schools, colleges, and educational assessment programs to stop using computer scoring of student essays written during high-stakes tests."

As of early 2015, the petition has over 4,000 signatures. Including academics from institutions of all sizes, many of them the recognized players in the literacy education world. Many public schoolteachers. A corporate attorney. Noam Chomsky and Professor Perelman representing MIT. Clearly, Haswell and his collaborators were onto something—there are a lot of us out there who want to make our voices heard.

But it's important to look closely at what we're saying: "stop using computer scoring of student essays written during high-stakes tests." The guilty party, yet again, is the computer. The fact that human scorers have demonstrated themselves to be just as bad gets ignored. And the real culprits—the "essays" (read: papers) and "high-stakes tests" and the mechanized approach to teaching that makes them possible—don't even get indicted.

Why? Because they're just the way things are. It's not that we like them; we hate them, but there's nothing we can do about them right now. They've become so natural, so normal, that even their most dedicated opponents can't imagine fighting them.

Likewise, alas, with the formal position statement from our colleagues at CCCC:

> *Writing-to-a-machine violates the essentially social nature of writing: we write to others for social purposes. If a student's first writing-experience at an institution is writing to a machine, for instance, this sends a message: writing at this institution is not valued as human communication.*[31]

When we read this, even now for the umpteenth time, we find ourselves simultaneously wanting to cheer and cringe. Cheer, because we're proud that our national organization is promoting the cause we've dedicated our lives to: teaching writing and literacy as something social, as a form of human communication, which brings people together and gives them a voice. Cringe, because it's naive and wrongheaded. Education in the United States is not social, human, or empowering, and we can't blame that on the machine.

Or in the words of Paul Deane, the ETS representative who faced off against Les in St. Louis:

> *Current-generation AES systems largely presuppose the definition of the writing construct implicit in a standardized testing situation and embed automated scoring within that frame . . . objections to AES are actually objections to the assumptions of standardized testing . . . the CCCC position statement . . . claims that AES sends the message that writing is not valued primarily*

as a form of communication; but the standardized situation has already changed the audience and the purpose of writing.[32]

Deane is right. The problem isn't the technology, but the nature of the "writing construct." We made it what it is. They just dutifully (and lucratively) responded to it.

So, can we change it? And if so, how?

WHY THEY'RE STILL WRONG

First, take a look at these two pieces of student writing. Number One:

In my younger mindset, I never doubted that God has a plan for everyone, and that all things are in his hand . . . until Simon died. I still believe that God is omni-powerful, but I also think that he granted the world free will, and with this gift, we are to make our own choices in an accident-prone life. Once Simon's accident happened, I couldn't bring myself to believe that God planned for a 19-year-old child and three of his friends to be killed by two drunk drivers on the road. I do not believe that Simon, his friends, or the drivers they collided into, were fulfilling some sort of divine path, and I refuse to attribute this happening or any other depressing situation occurring in the world to God's or any other Spirit's plan. Hegel's idea of a World Spirit driving the occurrences in reality to pave the way for a specific plan fulfillment is similarly difficult for me to believe.

And Number Two:

There exist many cultural inequalities in society that modern feminists often argue about. Examples include: sex and gender being socially constructed, and that traditional female roles in society are too limiting. Furthermore, popular culture has often given women degrading roles in films and music. Women are stereotypically portrayed as the damsel in distress in movies or sex objects in songs. Heavily made-up women are only to be

> *seen on the big screen, and only prove important if attractive. Furthermore, women are too often viewed as sex objects; their roles in popular media are predominantly sexual. Men are rarely ever seen as merely sex objects. What society perceives as normal is far from what feminists would hope for in an equal world.*

Believe it or not, both of these pieces were written by the same student. In the same semester, in the same course. The only difference was the assignment. The assignment that produced Number Two was, basically, *write a "paper."* The assignment that produced Number One was *write an email to someone close to you, telling that person about something we've read and why s/he might find it important—and then actually send it to that person*.

Why so different? Each assignment called up a very different "writing construct." Number Two comes right out of what Deane called the "standardized" construct. e-Rater®, or its human equivalents, would probably love it. Big words, used not entirely correctly. Long, unnecessarily complex sentences. Two "furthermores" in a single paragraph. And of course, it doesn't say much; it's fake. This is the "paper." It's the "shit my students write."

The bad news: it's also the standard. By the time students get to college, this is how they know how to write—their dispositions are pretty well formed. And at the K–12 level, where more could be done, teachers are usually too constrained to do it. In many schools, especially the poorer ones, they have no choice but to spend months each year preparing for standardized tests, including teaching the standardized writing construct. ("According to the text . . .") Then there's the weeks (if not months) of instructional time they lose in administering these tests. And on top of all that, they're often contending with huge classes and course loads, and minimal resources—hardly a great environment for the kind of labor-intensive teaching needed to re-form a student's dispositions into a better writing construct.

(Nor is ETS quite as innocent as Deane claims. Education technology is less a free market than an oligopoly, where a few heavy hitters set the rules for everyone else. ETS spends immense amounts of time and money convincing educators and policymakers that standardized testing is a good idea. They may not have invented the standardized construct, but they do a lot to keep it in place. In some ways they behave better than their for-profit counterparts—they allowed Les to test their AES software, when Pearson and Vantage refused. At the same time, their 501(c)(3) nonprofit status and veneer of academic legitimacy give them a public credibility that the others lack—which they use less to serve the public good than to advance their industry.)[33]

The good news: the "standardized" construct can be broken. Witness Number One. ETS would probably give it a lower score. It's a much simpler piece of writing, at least in the ways they'd count. But in every way that counts, it's a better piece of writing. It makes a much more complex and interesting claim. It works through hard texts and difficult ideas, in a way that's genuine and honest. It connects school to real life. Above all, it's personal and meaningful; it's real.

And breaking the standardized construct was easier than we might think. All it took was a little critical thought about the way we teach and the constructs we're working with. When we go on autopilot and just say "write a paper" (or even "write about x"), we'll get the standardized construct. But if we invest the time and labor into setting up the construct intentionally, we can get something much better. Something not just graded by humans but actually human.

There are a lot of good examples out there, if we know where to look. And a great many of them involve a lot of advanced technology.

Wikipedia, for many teachers, is a problem. Students use it to avoid doing "actual" research, and it's not uncommon to find plagiarized paragraphs cut-and-pasted wholesale into their papers. But we can't just ban Wikipedia. They'll keep using it anyway. (And if we're honest, so will we.) The better answer is the one proposed by our colleagues Michael Kuhne and Gill Creel at Minneapolis Community and Technical College: teach with it.

In their first-year writing course, students are *required* to use Wikipedia: to read it, to write about it, and ultimately, to write *in* it. The final assignment of the course: get some text up onto Wikipedia that stays there for a while. This is harder than it sounds. Try to edit a Wikipedia entry, and you'll face an onslaught of editors, both humans and Bots. The Bots work a lot like AES: they evaluate edits based on formal features, and they instantly decide what stays and what goes. And part of the course—in discussions, and in reflective writing—is about learning to understand the Bots, what they're looking for, and how to write for them.[34]

This kind of teaching is saturated in technology, but there's nothing fake about it. They're writing about topics they care about (they choose which articles to edit), and they're writing for an audience of millions of Wikipedia-readers.

It's important to note that Michael and Gill teach in a two-year college with massive teaching loads. They know as well as anyone that writing ability comes from labor, and they know they can't possibly provide minute feedback to each of the 120-some students they meet. But Wikipedia can. Not as an artificial labor-saving device a la AES, but as a real-world digital tool, just like the spellcheck function in Google Docs that just put a red line under the word "spellcheck." (Yes, we wrote this book in Google Docs, collaborating long-distance. And we couldn't have done it without the open-source scholarship provided by Utah State University Digital Commons.)

Carl Whithaus, director of the University Writing Program at UC Davis—the guy who sat between Les and Deane at the CCCC panel—makes this same point: "Outright rejection of software as an assessment and response tool is not a viable, practical stand, because software is already reading, responding, and assessing student writing."[35]

Whithaus provides lots of examples of technology-driven "multimodal" teaching that speaks directly to how real people write today. He can sometimes overlook the distinction between "real" and "fake" technology, between technology used to break the standardized construct versus to maintain it, but his call to embrace the real world is refreshing.[36]

For teachers like us, who are a little too nervous to try something as ambitious as Kuhne and Creel's Wikipedia project, Whithaus and his colleagues also offer tons of simpler possibilities—some digital, others not. Take student writing Number One, from the start of this section. It's a long-form analytical writing project, in many ways like a paper. But with two big differences: it's an email (a real-life form of writing) to her father (a real-live audience).

And he got it—and wrote back:

> *Your arguments in the message are logical and coherent. God does not create us like [robots], but like image[s] of God with full freedom. A notion of freedom is tempting when it is dissociated from God's grace. God's grace does not destroy our freedom, but completes it. Let's talk more about the complex thought of Hegel's philosophy of Geist later. Cordially, Your daddy.*

Such assessment, no machine can provide.

WHAT WE CAN DO ABOUT IT

We know what needs to happen. It won't be easy—by definition, it will be extremely labor-intensive—but if we have the will, the time, and the power to take back some control over the curriculum and the debate over education technology, we can do it.

Of course, getting those things will be a challenge. Especially that last one. Witness Walter Stroup, professor at the University of Texas. He published a study that showed exactly what we have been claiming: that his state's high-stakes tests did not measure learning so much as class-based dispositions.

In response, the Pearson Center for Applied Psychometric Research, a multimillion-dollar institute on the UT campus backed by the same corporate behemoth that supplies the tests, kicked into high gear. A slew of studies appeared, filled with statistics, seemingly overnight. They couldn't conclusively prove Stroup wrong, but they cast lots of doubt on his work. Before long, he found his entire academic career was in jeopardy. And of course, the test remains.[37]

This is what we're up against. *They* are well organized and extremely well resourced, and *we* are not. Stories like Stroup's could convince anyone it's hopeless. Yet Michael Gecan, one of the country's most experienced community organizers, insists otherwise:

> *Wherever I go in higher education circles, I hear a great deal of anxiety and concern about the pressures that faculty and other staff members are facing. Budget cuts keep coming, particularly for community colleges, but across the board for public institutions. The reduction of full-time positions in favor of adjunct, part-time, and contracted-out services continues. The latest "new-new thing," Massive Open Online Courses (MOOCs), is being pushed and promoted by elite universities on both coasts and is funded by the Silicon Valley crowd. And the transformation of*

> *higher education from its former role as an engine of new opportunity and prosperity for people into a sinkhole of unbearable debt seems relentless and perhaps irreversible to many.*
>
> *Yet, as real as these pressures are and as overwhelming as they can seem, they can be faced—and reduced or reversed—by well-organized people who know how to play a better brand of defense in the public arena, but who also develop a creative and productive offense. I say this not to diminish the reality and the difficulty of the issues in higher education, but to put that reality and those difficulties into some perspective. And I say this because I have seen many other leaders, in many other places, at times and in circumstances that seemed at least as dire, demonstrate that it could be done.*[38]

"Well-organized people who know how to play a better brand of defense" and a "creative and productive offense." What would that look like?

We can't provide a cookbook. Effective organizing, by nature, is locally-driven, with the people on the ground setting the agenda. But we can provide some general direction. For starters, we know what an effective campaign *can't* look like. It can't look like a fight against the machines. We won't win. And even if we do, it won't solve the real problem.

And it can't look like a "vigorous but civil debate" at an academic conference. Les has been tilting at ETS for over a decade. It's gotten attention, it's gotten press, and it's gotten the essay removed from the SAT. But it hasn't made a dent in the real problem.

And it can't look like a position statement or a petition. They feel good to write and sign on to, especially for those of us with academic dispositions who like well-cited arguments and like to be right, but they go nowhere. (Literally: the CCCC statement and the Human Raters petition have just sat there on the Internet, for years.)

But in one way at least, all these efforts have been successful. They have proven beyond a doubt—Les with his frequent appearances across the country and in the *Times*, Haswell with his 4,000+ signatures—that there are people out there who care. Who refuse to accept that There Is No Alternative to mechanized education. Who can be organized.

Into a zombie horde, for instance. No, seriously. In 2013, "the Providence [Rhode Island] Student Union, a group of student activists, who convened initially to protest [a high-stakes standardized test] as a graduation requirement, staged a protest march to the RI Department of Education dressed as zombies." A participant explained: "We're zombies because this policy will kill us."[39] A little silly perhaps, but what the students did next definitely wasn't: they "organized a group of prominent adults to take the state graduation test, a majority of whom failed the math."[40]

Nor were they alone. During the year 2013, according to the national network FairTest:

> *Students, parents and teachers engaged in boycotts, 'opt-out' campaigns, and walkouts in Seattle, Portland OR, Denver, Chicago and New York, with smaller events in other communities across the nation. Seattle teachers in several schools boycotted one test; most students, backed by their parents, did not take it when administrators gave it. Families boycotted at 40 schools in New York City and many outside the City. These direct actions generally attracted significant media attention. The goals varied across the groups, from a desire to roll back the amount of standardized testing and end high stakes, to stopping or reducing specific exams, such as for graduation. In other instances, goals were less defined or included non-testing issues. Some groups established long-term and interim goals, with the latter aimed at both useful victories and building capacity for further changes. For many groups, the most important outcome this year*

> *was enhanced capacity. Interviewees regularly talked about how much more community awareness they see, how many more people are involved in the effort.*[41]

And FairTest is only one network. The unions, including the National Education Association (who gave Gecan a platform to talk to teachers) and the Service Employees International Union (who is actively organizing educators and staff at the grade-school and college level, including Robin), are another. Likewise with groups like the American Association of University Professors (with their new Campaign for the Future of Higher Education), Imagining America (a national organization of educators and artists, including Ben, working to bring back education's public purpose), and maybe even the Tea Party (they may take a lot of positions we don't like, but on the issue of bureaucrats controlling their kids' education, they're right on).

We may not like all these people. We will certainly not always agree with them: their goals, their tactics, their selection of issues, some of their positions. It will be tempting to "problematize" (a.k.a. denounce) them and go our separate way.

But if we're smart, we won't. If we're serious about building the power necessary to challenge the immense power of Pearson, ETS, and the bureaucrats, we'll work together anyway. (In the immortal words of Saul Alinsky: "No permanent friends, no permanent enemies.") And that power, as Gecan explains, will only come in two forms: "organized people and organized money."[42]

From the looks of it, there are a lot of people and pocketbooks out there, waiting to be organized. Apart from a small, powerful group of corporations and government bureaucrats (of both parties), most people don't like mechanized schooling. They don't like standardized tests taking over the curriculum. They don't like MOOCs and robo-graders replacing teachers.

They don't like tech startups privatizing education. Even after decades of well-orchestrated bad press, as education expert Diane Ravitch reports, most people still support their local public schools.[43]

The people, it seems, are on our side. The problem, daunting as it may be, is ours to solve.

Notes

1. Les Perelman, "Essay Awarded a Top Grade by e-Rater," accessed September 24, 2014, http://www.documentcloud.org/documents/346138-essay-awarded-a-top-grade-by-e-Rater.html.

2. Michael Winerip, "Facing a Robo-Grader? Just Keep Obfuscating Mellifluously," *New York Times*, April 22, 2012, accessed September 24, 2014, http://www.nytimes.com/2012/04/23/education/robo-readers-used-to-grade-test-essays.html?pagewanted=all&_r=0.

3. Michael Winerip, "SAT Essay Test Rewards Length and Ignores Errors," *New York Times*, May 4, 2005, accessed September 24, 2014, http://www.nytimes.com/2005/05/04/education/04education.html.

4. The words of Lynn Streeter, from Knowledge Analysis Technology, promoting their Intelligent Essay Assessor. See Chris Anson, "Can't Touch This: Reflections on the Servitude of Computers as Readers," in *Machine Scoring of Student Essays: Truth and Consequences*, ed. Patricia Freitag Ericsson and Richard H. Haswell (Logan: Utah State University Press, 2006), 38–56.

5. Patricia Freitag Ericsson, "The Meaning of Meaning: Is a Paragraph More Than an Equation?" in *Machine Scoring of Student Essays*, ed. Ericsson and Haswell, 28–37.

6. Randy Elliot Bennett, "Automated Scoring of Constructed-Response Literacy and Mathematics Items," accessed September 27, 2014, http://www.ets.org/s/k12/pdf/k12_commonassess_automated_scoring_math.pdf.

7. "May 2014—SAT Essay Prompts," accessed November 6, 2015, http://www.cracksat.net/sat/essay/31.html.

8. "Essay Scoring: How It's Scored, and What the Scores Mean," accessed September 27, 2014, http://sat.collegeboard.org/scores/sat-essay-scoring-guide. Here's the full text of ETS's scoring rubrics to get a 6, the highest score. The differences between the six levels are differentiated only by "valuative" criteria—expressed as adjectives like "effectively," "severely flawed," or "meaningful":

- Effectively and insightfully develops a point of view on the issue and demonstrates outstanding critical thinking, using clearly appropriate examples, reasons and other evidence to support its position
- Is well organized and clearly focused, demonstrating clear coherence and smooth progression of ideas
- Exhibits skillful use of language, using a varied, accurate and apt vocabulary
- Demonstrates meaningful variety in sentence structure
- Is free of most errors in grammar, usage and mechanics

9. Professor Richard Haswell likewise wonders "why these few particular traits were chosen out of the plentiful supply good writers utilize, such as wit, humor, surprise, originality, logical reasoning, and so on." (He goes on to propose an answer both convincing and disturbing, which we'll discuss later.) See Richard H. Haswell, "Automatons and Automated Scoring: Drudges, Black Boxes, and Dei Ex Machina," in "Machine Scoring of Student Essays," ed. Ericsson and Haswell, 57–79.

10. Winerip, "SAT Essay Test Rewards Length and Ignores Errors."

11. John Markoff, "Essay-Grading Software Offers Professors a Break," *New York Times*, April 4, 2013, accessed April 8, 2013, http://www.nytimes.com/2013/04/05/science/new-test-for-computers-grading-essays-at-college-level.html?hp&_r=3&pagewanted=all&.

12. Ibid.

13. Illinois State University professor Bob Broad is one of few scholars who have given this matter serious thought. He shows that talk of "giving teachers a break" does in fact up from time to time in the world of AES promotion—such as the ETS postcard that promised their AES software "Criterion™ gives teachers what they need most . . . time to teach." Of course, Broad observes, "it is highly unlikely that outsourcing evaluation to ETS will result in teachers having more time to teach writing. To the contrary, writing teachers in institutions that purchase Criterion would almost surely be assigned more students or more classes, or both." See Bob Broad, "More Work for Teacher? Possible Futures for Teaching Writing in the Age of Computerized Writing Assessment," in "Machine Scoring of Student Essays," ed. Ericsson and Haswell, 221–33.

14. Former US Assistant Secretary of Education Diane Ravitch lays this out plainly: "We know what works. What works are the very opportunities that advantaged families provide for their children. In homes with adequate resources, children get advantages that enable them to arrive in school healthy and ready to learn. Discerning, affluent parents demand schools with full curricula, experienced staffs, rich programs in the arts, libraries, well-maintained campuses, and small classes. As a society, we must do whatever is necessary to extend the same advantages to children who do not have them. Doing so will improve their ability to learn, enhance

their chances for a good life, and strengthen our society." See Diane Ravitch, *Reign of Error: The Hoax of the Privatization Movement and the Danger to America's Public Schools*, reprint ed. (New York: Vintage, 2014), 6.

15. Sociolinguist Shirley Brice Heath (personal communication) set out to find out what made literacy instruction "work" and found out that there was almost no research out there. Of the evidence that existed, it seemed the most effective interventions served to give participants a concrete, relevant, personally-important image of themselves literate: something real, something that they could see changing their lives. Importantly, most of these interventions were not explicitly focused on literacy: mural-painting clubs, dance lines, drum corps, Bible-study groups, and Boys and Girls Clubs.

16. For more on cultural capital and durable dispositions, see Pierre Bourdieu, *Language and Symbolic Power*, ed. John B. Thompson, trans. Gino Raymond and Matthew Adamson (Cambridge, MA: Harvard University Press, 1991), especially chapters 2 and 4.

17. Bourdieu uses "field" and "market" interchangeably. We'll use "market" here, because we think it's clearer.

18. Les Perelman, "Dr. Perelman's SAT Writing Tips," accessed November 6, 2015, http://www.actoutagainstsat.com/essay-tips.pdf.

19. Winerip, "Facing a Robo-Grader?"

20. For this and many more MOOC-related statistics, see Katy Jordan, "The MOOC Project," accessed September 26, 2014, http://www.katyjordan.com/MOOCproject.html. Conventional, limited-enrollment online courses do better, but not by much. Our colleagues who teach them report that the dropout rate is about 50 percent.

21. See Rebecca Schuman, "The King of MOOCs Abdicates the Throne," *Slate Magazine*, November 19, 2013, accessed September 26, 2014, http://www.slate.com/articles/life/education/2013/11/sebastian_thrun_and_udacity_distance_learning_is_unsuccessful_for_most_students.html.

22. Professors Anne Herrington and Charles Moran observe this same "two-class system" across the country: "On the list of institutions using [AES programs] WritePlacer Plus or Criterion there is no Harvard or Princeton, no Williams or Oberlin or Amherst. There is, however, Truckee Meadows Community College, Camden County Community College, University College of the Cariboo, the University of South Florida, Northern Arizona University, and Valley College. The distribution of this product suggests to us an extension of the social and economic stratification that has been such a feature of the past decade: the wealthy and connected learn to write to make meaning and to achieve their rhetorical purposes; the poor and unconnected learn to write to scoring engines. We don't think that

this two-class system is the conscious aim of our institutions or of the people who administer and teach in them. But it may be the result of incremental decisions—to use computers to score placement essays, and then to give feedback to writers in writing centers, and then to use them to read exit exams for 'value added,' and then to grade papers in a large lecture course—each decision not made in a vacuum but in an atmosphere created by heavy teaching loads, underfunded public institutions, heavy marketing, and claims of 'efficiency' and 'authenticity.'" See Anne Herrington and Charles Moran, "WritePlacer *Plus* in Place: An Exploratory Case Study," in "Machine Scoring of Student Essays" ed. Ericsson and Haswell, 114–30.

23. As professors Ken S. McAllister and Edward M. White describe: "Some teachers and administrators turn reflexively to technological solutions when funding for human labor is in crisis, as has been the case for education at all levels in recent years. They gain support from those who turn to technological solutions for other reasons, such as a genuine interest in new methodologies as well as the novelty and 'coolness' factors they bring." See Ken S. McAllister and Edward M. White, "Interested Complicities: The Dialectic of Computer-Assisted Writing Assessment," in "Machine Scoring of Student Essays," ed. Ericsson and Haswell, 8–27.

24. Ericsson and Haswell note that "on the issue of automated essay scoring teachers and students have not been encouraged to ask questions at all and sometimes have been systematically excluded from forums where their opinion should have a voice and might have an appeal. In *The Neglected "R": The Need for a Writing Revolution* (2003), authored by the National Commission on Writing in America's Schools and Colleges and published by the College Board [a.k.a. ETS], teachers and students are excluded from the agenda for 'revolution.' Instead it is recommended that "the private sector work with curriculum specialists, assessment experts, and state and local educational agencies to apply emerging technologies to the teaching, development, grading, and assessment of writing." In many cases even school administrators are excluded, such as in the Texas State Board of Education's decision to adopt AES on the statewide college entrance examination—a decision "made by state government officials and industry lobbyists with no input from writing experts or administrators in higher education." This teacherless decision-making process is facilitated by the rapid growth of what we might call "para-academic" knowledge centers within education technology companies. ETS, for example, employs well-credentialed researchers with named "chairs" and academic titles, who produce lots of papers reporting solid and expensive research findings, often peer-reviewed in journals supported by the industry, and populated with their people. When the marketing department wants to push a new machine or service, or when the sales force encounters a room full of principals, superintendents, and school-board members, it's easy to say "research supports." See Patricia Freitag Ericsson and Richard H. Haswell, "Introduction," in *Machine Scoring of Student Essays*, ed. Ericsson and Haswell, 1–7.

25. See Broad, "More Work for Teacher?," 223.

26. Haswell elaborates: "With our decades-long trust in holistic scoring, we have again already bought into machine scoring . . . machine scoring and holistic scoring enjoy a relationship that is historically complementary, even mutually supportive, maybe even symbiotic. . . . The traits e-Rater uses have a long history with essay assessment, and in particular with holistic scoring at Educational Testing Service. History is the trial that shows us these traits are especially important to writing teachers." See Haswell, "Automatons and Automated Scoring," 68–70.

27. "A 'Whatever' Type of Thing," *Shit My Students Write*, February 2, 2015, accessed November 6, 2015, http://shitmystudentswrite.tumblr.com/post/109889919239/a-whatever-type-of-thing; "Not Your Stereotypical Artist," *Shit My Students Write*, August 6, 2014, accessed November 6, 2015, http://shitmystudentswrite.tumblr.com/post/94033066089/not-your-stereotypical-artist; "Inherently Venereal," *Shit My Students Write*, July 19, 2014, accessed November 6, 2015, http://shitmystudentswrite.tumblr.com/post/92278264529/inherently-venereal; "The March of Time," *Shit My Students Write*, July 6, 2014, accessed November 6, 2015, http://shitmystudentswrite.tumblr.com/post/90969225564/the-march-of-time.

28. Haswell, "Automatons and Automated Scoring," 59.

29. Broad claims, not without reason, that "a more candid motto for [ETS's AES software] Criterion would be: 'Teach your students to write like machines for a reader who is a machine.'" But as Haswell notes, we have been teaching our students to write like machines, however unwittingly, since long before ETS came on the scene. See Broad, "More Work for Teacher?" 224.

30. Haswell, "Automatons and Automated Scoring," 65. Haswell and Ericsson elaborate: aside from a few books and "the occasional sparsely attended conference presentation, the response on machine scoring from the academic community, especially from writing teachers and composition scholars, has been silence." See Haswell and Ericsson, "Introduction," 2.

31. "CCCC Position Statement on Teaching, Learning, and Assessing Writing in Digital Environments," *Conference on College Composition and Communication*, February 2004, accessed September 9, 2014, http://www.ncte.org/cccc/resources/positions/digitalenvironments.

32. Paul Deane, "On the Relation Between Automated Essay Scoring and Modern Views of the Writing Construct," *Assessing Writing* 18, no. 1 (January 2013): 7–24.

33. It's also worth noting that ETS is not entirely not-for-profit. ETS Technologies, Inc. is "a for-profit, wholly-owned subsidiary of ETS." See Broad, "More Work for Teacher?," 222.

34. See Michael Kuhne and Gill Creel, "Wikipedia, 'the People Formerly Known as the Audience,' and First-Year Writing," *Teaching English in the Two-Year College* 40, no. 2 (December 2012): 177–89.

35. Whithaus goes on to discuss interesting ideas for crafting assignments using the Microsoft Word spellchecker. "By working with the features in Microsoft Word such as readability and by explaining to students exactly how the grammar-checker works in terms of their language, we make the students' interaction with the software into teachable moments rather than rote acceptance of the software's authority." See Carl Whithaus, "Always Already: Automated Essay Scoring and Grammar-Checkers in College Writing Courses," in *Machine Scoring of Student Essays*, ed. Ericsson and Haswell, 166–76.

36. To see the wide scope of his work, see Carl Whithaus, *Teaching and Evaluating Writing in the Age of Computers and High-Stakes Testing* (Mahwah, NJ: Lawrence Erlbaum Associates, 2005).

37. See Jason Stanford, "Mute the Messenger," *Texas Observer*, September 3, 2014, accessed February 9, 2015, http://www.texasobserver.org/walter-stroup-standardized-testing-pearson/.

38. Michael Gecan, "Building Better Relationships for Better Organizing," *Thought & Action* (Fall 2013): 27–34, 28, accessed September 26, 2014, http://www.nea.org/assets/docs/HE/TA2013Gecan.pdf.

39. Diane Elliot Kern, "Zombie Ideas in Education: High-Stakes Testing and Graduation Policies," *New England Reading Association Journal* 49, no. 1 (2013): 96.

40. FairTest, "The Spring 2013 Testing Reform Uprising," accessed February 9, 2015, http://www.fairtest.org/sites/default/files/The-Spring2013-Testing-Reform-Uprising.pdf.

41. Ibid.

42. Michael Gecan, *Going Public: An Organizer's Guide to Citizen Action* (New York: Anchor Books, 2002), 36.

43. Ravitch, *Reign of Error*, 39.

About the Authors

Ben Fink is a teacher, organizer, and consultant. He has directed programs in critical pedagogy, arts education, and community engagement in academia and the nonprofit sector. He taught many years of writing-intensive courses at the University of Minnesota and is now Creative Placemaking Project Manager at Appalshop, Inc.

Robin Brown is Morse-Alumni Distinguished Professor at the University of Minnesota in the cultural studies department, where he teaches science studies and pedagogy. His career has always been interdisciplinary and recently has focused on the multiple relationships of rhetoric, science, technology, politics, and identity.

Current Arguments in Composition

Utah State University Press's Current Arguments in Composition is a series of short-form publications of provocative original material and selections from foundational titles by leading thinkers in the field. Perfect for the composition classroom as well as the professional collection, this series provides access to important introductory content as well as innovative new work intended to stimulate scholarly conversation. Volumes are available in paperback or ebook form.